Designing and Evaluating Symbols for Electronic Displays of Navigation Information: Symbol Stereotypes and Symbol-Feature Rules

Michelle Yeh

Divya C. Chandra

U.S. Department of Transportation

Research and Innovative Technology Administration

John A. Volpe National Transportation Systems Center

Cambridge, MA 02142

DOT/FAA/AR-05/48

DOT-VNTSC-FAA-05-16

September 2005

For:
Dr. Tom McCloy
Federal Aviation Administration (FAA)
Air Traffic Organization Operations Planning
Human Factors Research and Engineering
(ATO-P R&D)

This document is available to the public through the National Technical Information Service, Springfield, Virginia, 22161

Notice

This document is disseminated under the sponsorship of the Department of Transportation in the interest of information exchange. The United States Government assumes no liability for its contents or use thereof.

Notice

The United States Government does not endorse products or manufacturers. Trade or manufacturers' names appear herein solely because they are considered essential to the objective of this report.

REPORT DOCUMENTATION PAGE			Form Approved OMB No. 0704-0188
Public reporting burden for this collection of information is estimated to average 1 hour per response, including the time for reviewing instructions, searching existing data sources, gathering and maintaining the data needed, and completing and reviewing the collection of information. Send comments regarding this burden estimate or any other aspect of this collection of information, including suggestions for reducing this burden, to Washington Headquarters Services, Directorate for Information Operations and Reports, 1215 Jefferson Davis Highway, Suite 1204, Arlington, VA 22202-4302, and to the Office of Management and Budget, Paperwork Reduction Project (0704-0188), Washington, DC 20503.			
1. AGENCY USE ONLY (Leave blank)	2. REPORT DATE September 2005	3. REPORT TYPE AND DATES COVERED Final Report, September 2005	
4. TITLE AND SUBTITLE Designing and Evaluating Symbols for Electronic Displays of Navigation Information: Symbol Stereotypes and Symbol-Feature Rules		5. FUNDING NUMBERS FAE2 BB079 FA6Y BD305	
6. AUTHOR(S) Michelle Yeh and Divya C. Chandra			
7. PERFORMING ORGANIZATION NAME(S) AND ADDRESS(ES) U.S. Department of Transportation John A. Volpe National Transportation Systems Center Research and Innovative Technology Administration Cambridge, MA 02142-1093		8. PERFORMING ORGANIZATION REPORT NUMBER DOT-VNTSC-FAA-05-16	
9. SPONSORING/MONITORING AGENCY NAME(S) AND ADDRESS(ES) U.S. Department of Transportation Federal Aviation Administration Office of Aviation Research, Human Factors Research and Engineering Division 800 Independence Avenue, SW Washington, D.C. 20591 Program Manager: Dr. Tom McCloy		10. SPONSORING/MONITORING AGENCY REPORT NUMBER DOT/FAA/AR-05/48	
11. SUPPLEMENTARY NOTES			
12a. DISTRIBUTION/AVAILABILITY STATEMENT This document is available to the public through the National Technical Information Service, Springfield, VA 22161		12b. DISTRIBUTION CODE	
13. ABSTRACT (Maximum 200 words) There is currently no common symbology standard for the electronic display of navigation information. The wide range of display technology and the different functions these displays support makes it difficult to design symbols that are easily recognizable across platforms. The goals of this effort are to identify features of navigation symbology that are problematic when presented on electronic displays and to develop a method to design and evaluate symbology that takes into account the different media and platforms on which they will be displayed. This report presents the findings of two studies related to the design of symbology for electronic displays of navigation information. Experiment 1 addressed whether symbols have key features that are necessary for recognition, i.e., *symbol stereotypes*. Experiment 2 tested *symbol-feature rules*, which define a consistent way to design symbols, to determine if pilots could learn and apply them. The resulting guidelines were provided to the Federal Aviation Administration, industry, and the International Civil Aviation Organization and are documented in this report. Suggestions for evaluating symbology are also provided.			
14. SUBJECT TERM Aeronautical chart symbology, electronic displays, moving map display, navigation displays, symbol design, map display		15. NUMBER OF PAGES 51	
		16. PRICE CODE	
17. SECURITY CLASSIFICATION OF REPORT Unclassified	18. SECURITY CLASSIFICATION OF THIS PAGE Unclassified	19. SECURITY CLASSIFICATION OF ABSTRACT Unclassified	20. LIMITATION OF ABSTRACT

METRIC/ENGLISH CONVERSION FACTORS

ENGLISH TO METRIC

LENGTH (APPROXIMATE)
- 1 inch (in) = 2.5 centimeters (cm)
- 1 foot (ft) = 30 centimeters (cm)
- 1 yard (yd) = 0.9 meter (m)
- 1 mile (mi) = 1.6 kilometers (km)

AREA (APPROXIMATE)
- 1 square inch (sq in, in^2) = 6.5 square centimeters (cm^2)
- 1 square foot (sq ft, ft^2) = 0.09 square meter (m^2)
- 1 square yard (sq yd, yd^2) = 0.8 square meter (m^2)
- 1 square mile (sq mi, mi^2) = 2.6 square kilometers (km^2)
- 1 acre = 0.4 hectare (he) = 4,000 square meters (m^2)

MASS - WEIGHT (APPROXIMATE)
- 1 ounce (oz) = 28 grams (gm)
- 1 pound (lb) = 0.45 kilogram (kg)
- 1 short ton = 2,000 pounds (lb) = 0.9 tonne (t)

VOLUME (APPROXIMATE)
- 1 teaspoon (tsp) = 5 milliliters (ml)
- 1 tablespoon (tbsp) = 15 milliliters (ml)
- 1 fluid ounce (fl oz) = 30 milliliters (ml)
- 1 cup (c) = 0.24 liter (l)
- 1 pint (pt) = 0.47 liter (l)
- 1 quart (qt) = 0.96 liter (l)
- 1 gallon (gal) = 3.8 liters (l)
- 1 cubic foot (cu ft, ft^3) = 0.03 cubic meter (m^3)
- 1 cubic yard (cu yd, yd^3) = 0.76 cubic meter (m^3)

TEMPERATURE (EXACT)
[(x-32)(5/9)] °F = y °C

METRIC TO ENGLISH

LENGTH (APPROXIMATE)
- 1 millimeter (mm) = 0.04 inch (in)
- 1 centimeter (cm) = 0.4 inch (in)
- 1 meter (m) = 3.3 feet (ft)
- 1 meter (m) = 1.1 yards (yd)
- 1 kilometer (km) = 0.6 mile (mi)

AREA (APPROXIMATE)
- 1 square centimeter (cm^2) = 0.16 square inch (sq in, in^2)
- 1 square meter (m^2) = 1.2 square yards (sq yd, yd^2)
- 1 square kilometer (km^2) = 0.4 square mile (sq mi, mi^2)
- 10,000 square meters (m^2) = 1 hectare (ha) = 2.5 acres

MASS - WEIGHT (APPROXIMATE)
- 1 gram (gm) = 0.036 ounce (oz)
- 1 kilogram (kg) = 2.2 pounds (lb)
- 1 tonne (t) = 1,000 kilograms (kg)
- = 1.1 short tons

VOLUME (APPROXIMATE)
- 1 milliliter (ml) = 0.03 fluid ounce (fl oz)
- 1 liter (l) = 2.1 pints (pt)
- 1 liter (l) = 1.06 quarts (qt)
- 1 liter (l) = 0.26 gallon (gal)
- 1 cubic meter (m^3) = 36 cubic feet (cu ft, ft^3)
- 1 cubic meter (m^3) = 1.3 cubic yards (cu yd, yd^3)

TEMPERATURE (EXACT)
[(9/5) y + 32] °C = x °F

QUICK INCH - CENTIMETER LENGTH CONVERSION

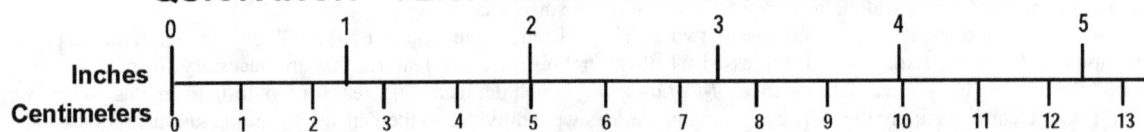

QUICK FAHRENHEIT - CELSIUS TEMPERATURE CONVERSION

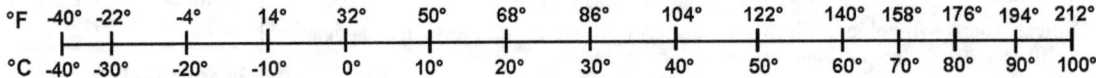

For more exact and or other conversion factors, see NIST Miscellaneous Publication 286, Units of Weights and Measures. Price $2.50

PREFACE

This report was prepared by the Operator Performance and Safety Analysis Division of the Office of Safety and Security at the Volpe National Transportation Systems Center. It was completed under the Division's Flight Deck Human Factors program and the Flight Deck Technologies program with funding from the Federal Aviation Administration's (FAA) Human Factors Research and Engineering Division (ATO-P R&D), in support of the Office of Aircraft Certification (AIR). We would especially like to thank our FAA Program Manager, Tom McCloy, as well as our Technical Sponsor, Colleen Donovan. Many thanks also to Bill Kaliardos, John Moore, Eric Secretan, and the many other FAA staff who have given us feedback and suggestions. Finally, thanks to Honeywell, Jeppesen, Rockwell Collins, and other manufacturers and chart providers who provided their symbol sets for use in our study; to Bill Edmunds, Andrew Kendra, Lt. Col. Wes Olson, and Doug White for their assistance recruiting participants for the studies; and to the participants for their time.

The results of the studies presented in this report address navigation symbology, although the techniques used here are applicable to addressing other types of symbology. The views expressed herein are those of the authors and do not necessarily reflect the views of the Volpe National Transportation Systems Center, the Research and Innovative Technology Administration, or the United States Department of Transportation.

Feedback on this document can be sent to Michelle Yeh (Yeh@volpe.dot.gov) or Divya Chandra (Chandra@volpe.dot.gov).

TABLE OF CONTENTS

List of Tables And Figures iv

Executive Summary v

1 Introduction 1

   1.1 Research Steps 1

   1.2 Research Issues 2

2 Experiment 1: Symbol Stereotypes 4

   2.1 Method 5

   2.2 Participants 5

      2.2.1 Symbols 5

      2.2.2 Tasks 5

      2.2.3 Procedure 6

   2.3 Data 7

   2.4 Results 7

      2.4.1 DME 7

      2.4.2 Fix 9

      2.4.3 NDB 10

      2.4.4 TACAN 11

      2.4.5 VOR 12

      2.4.6 VORDME 13

      2.4.7 VORTAC 14

      2.4.8 Waypoint 15

      2.4.9 Summary 16

3 Experiment 2: Symbol-Feature Rules 18

   3.1 Method 19

      3.1.1 Participants 19

      3.1.2 Symbols 19

      3.1.3 Task 20

      3.1.4 Procedure 22

   3.2 Data 22

   3.3 Results 22

      3.3.1 Fly-By vs. Fly-Over 22

      3.3.2 Compulsory vs. On-Request 24

      3.3.3 Ground-Based vs. RNAV 26

      3.3.4 Summary 28

4	Considerations and Steps for Evaluating Symbols	30
	4.1 Symbol Considerations	30
	4.2 Steps for Evaluating Symbols	31
5	Summary and Conclusion	33
6	References	34
Appendix A: Background Questionnaire		35
Appendix B: Symbol Stereotypes Questionnaire		37

LIST OF TABLES AND FIGURES

Table 1. Symbol-Feature Rule Summary. 2

Table 2. Variations in the VORTAC symbol. 4

Table 3. USA and ICAO fly-by and fly-over waypoint symbols in 1999. The ICAO symbol was changed in 2000. 4

Table 4. Symbol Recognition: DME. 8

Table 5. Symbol Recall: DME. 8

Table 6. Symbol Recognition: Fix. 9

Table 7. Symbol Recall: Fix. 9

Table 8. Symbol Recognition: NDB. 10

Table 9. Symbol Recall: NDB. 10

Table 10. Symbol Recognition: TACAN. 11

Table 11. Symbol Recall: TACAN. 11

Table 12. Symbol Recognition: VOR. 12

Table 13. Symbol Recall: VOR. 12

Table 14. Symbol Recognition: VORDME. 13

Table 15. Symbol Recall: VORDME. 13

Table 16. Symbol Recognition: VORTAC. 14

Table 17. Symbol Recall: VORTAC. 14

Table 18. Symbol Recognition: Waypoint. 15

Table 19. Symbol Recall: Waypoint. 15

Table 20. Symbol Stereotypes: Representative shape. 16

Table 21. Symbols defined by combining shape and fill. 18

Table 22. ICAO electronic symbols and their tested versions. 20

Table 23. Foil Example. 20

Table 24. Symbol-Feature Rules: Fly-By vs. Fly-Over. The NDB symbol was most problematic when applying the fly-by/fly-over symbol-feature rule to the paper symbols. 23

Table 25. Symbol-Feature Rules: Compulsory vs. On-Request. The NDB symbol was most problematic when applying the compulsory/on-request symbol-feature rule to the paper symbols. 25

Table 26. Symbol-Feature Rules: Ground-Based vs. RNAV. The fix symbol was most problematic when applying the ground-based/RNAV symbol-feature rule to the paper symbols. 27

Figure 1. Introductory slide: Minimal legend depicting the on-request vs. compulsory reporting point rule. 21

Figure 2. Introductory slide: Detailed legend depicting the on-request vs. compulsory reporting point rule. 21

EXECUTIVE SUMMARY

There is currently no common symbology standard for the electronic display of navigation information. The wide range of display technologies and the different functions these displays support make it difficult to design symbols that are easily recognizable across platforms. The Volpe Center worked with the Federal Aviation Administration (FAA) Office of Aircraft Certification (AIR), National Aeronautical Charting Office (NACO), and Human Factors Research and Engineering Division (ATO-P R& D) to identify and prioritize issues in symbology design, with the goal of supporting the development of symbol standards. The goals of the current effort were (1) to evaluate existing symbols and identify features of symbology that are problematic when presented on electronic displays and (2) to develop a method to design and evaluate symbology that takes into account the different media (e.g., paper vs. electronic) and platforms on which they will be displayed.

Two experiments are reported here. The first experiment addressed the issue of *symbol stereotypes*, i.e., whether there are key features that are necessary for symbols to be recognized. The second experiment tested symbol-feature rules to determine if pilots could learn and apply them. The rules provide a consistent way to construct navigation symbols so that a symbol conveys specific properties of the navigation aid, such as whether it is a fly-by or fly-over waypoint.

The results of the first experiment showed that pilots do have stereotypes regarding what symbol shapes are representative of a symbol type. Stereotypical shapes for navigation symbols were identified despite variations in the size, color, and fill with which the test symbol shapes were presented. The results of the second experiment showed that pilots were generally able to learn and apply the proposed symbol-feature rules, although pilots were better able to apply the rules when instruction was provided in a detailed legend that explicitly described the rule and depicted many examples as compared to a minimal legend that showed only a couple examples.

The results suggest that symbol stereotypes are a reality and should be considered in the design of electronic symbols to maintain safety. These results, taken together with recommendations for evaluating new symbology, support the development of recommended best practices (i.e., the Society of Automotive Engineers (SAE) G-10, Aerospace Behavioral Engineering Technology Committee, Aerospace Recommended Practice (ARP) 5289). This ARP is intended to be referenced in an update to FAA Advisory Circular 25-11 (Electronic Displays). While the scope of this work addresses navigation symbology, the techniques used here are applicable to addressing other types of symbology.

1 INTRODUCTION

An increasing number of electronic displays show navigation information, i.e., information from aeronautical charts that assists the pilot in determining the aircraft's position. Displays include in-flight moving map displays driven by a Flight Management System (FMS), electronic charts on an Electronic Flight Bag (EFB), surface moving map displays on an installed unit, or panel-mounted moving map displays on a Global Positioning System (GPS) unit. The design of symbology for navigation displays is complex due to this wide range of display technology and functionality.

Pilots must be able to extract and integrate information conveyed by symbols from electronic sources and paper charts for flight planning, situation awareness, and navigation. Often, pilots learn symbols shown on aircraft displays or paper charts and may form stereotypes regarding what shapes they expect to be used to represent various symbol types. These stereotypes facilitate recognition and interpretation, but may be problematic when transitioning between aircraft displays or between chart providers. In fact, inconsistencies in the symbols used on paper charts require air carriers to train their pilots on the differences in order to facilitate information retrieval and prevent misinterpretation. Thus, some level of commonality is important as pilots transition between aircrafts and display mediums (e.g., from paper charts to electronic map displays).

The goals of this effort are to identify features of navigation symbology that are problematic when presented on electronic displays and to develop a method to design and evaluate good symbology that takes into account the different platforms on which it will be displayed. Results of the research are expected to facilitate the design and evaluation of symbology, regardless of the display platform (i.e., paper or electronic).

1.1 Research Steps

In order to achieve our goals, we first needed to understand what display technologies and symbology are currently in use. Because a symbol's appearance on a display will vary depending on the physical qualities of the display, the first step was to determine the different displays in use. In September 2003, commercially available systems that included a moving map or chart display were identified through a web search and published product literature. These systems were primarily used by general aviation (GA) pilots. Because we were also interested in the capabilities of display technologies used in transport category airplanes, four transport category display manufacturers were contacted and asked to provide specifications for what they considered to be their low-end and high-end displays.

The results of the survey showed that from a symbology point of view, the biggest difference between low-end and high-end displays in the GA domain was resolution. For transport category displays, the results indicated that while display resolution was still important, it was not the major concern; in fact, the resolution of what was considered a low-end display for air transport aircraft was higher than the resolution for some high-end displays used in GA operations. Rather, the important issue was optimizing contrast, e.g., by increasing luminance or stroke width.

It was also important to determine what symbols were used on electronic displays. There is currently no standard set of symbology for the electronic display of navigation information. Several documents provide recommendations and guidelines for symbology for moving map displays (e.g., International Civil Aviation Organization (ICAO) standards document, *Annex 4, Aeronautical Charts* [2], and RTCA DO-257A, *Minimum Operational Performance Standards for the Depiction of Navigational Information on Electronic Maps* [5]). Additionally, the Society of Automotive Engineers (SAE) has developed an Aerospace Recommended Practice (ARP) for electronic aeronautical symbols [6]. However, informal discussions with manufacturers suggest that the SAE recommended symbols are not in widespread use. One reason is some of the symbols require a level of detail that is not possible to show on some displays. Another is that some of the recommended symbols are similar to copyrighted symbols; consequently, some manufacturers are wary about using them.

We contacted five aviation display manufacturers and asked them to send us their navigation symbol sets. In particular, we focused on eight navigation symbols: DME, intersection/fix, NDB, TACAN, VOR, VOR/DME, VORTAC, and waypoint. These eight symbols represent the majority of the navigation symbol types used in the United States of America (USA). Additionally, we collected symbols used on USA government FAA National Aeronautical Charting Office (NACO) charts [3] and those recommended in ICAO *Annex 4* [2] and SAE ARP 5289 [6]. We compared these eight symbol sets. The comparison highlighted the use of non-standard symbols and varying levels of detail in the symbols depending on the manufacturer. Thus, the potential for confusing and misleading symbology exists.

1.2 Research Issues

Based on our findings from the industry review and informal symbol comparison, a list of research issues that addressed how symbols are designed was compiled. The issues focused on factors that influence the legibility and comprehension of symbols. The issues were summarized into four basic questions for measuring the *usability* of a symbol (see Yeh and Chandra [9] for a discussion):

- Is the symbol shape representative of the symbol type?
- Can all encoded features of the symbol be decoded quickly and accurately?
- Is the symbol easy to find?
- Is the on-screen symbol size appropriate?

The Volpe Center worked with the FAA Office of Aircraft Certification (AIR), NACO, and Human Factors Research and Engineering Division (ATO-P R&D) to prioritize the research issues, with the goal of supporting efforts for developing symbol standards. One issue of concern was the variety of symbols shapes in use by different manufacturers. Additionally, symbol shapes used on electronic displays are sometimes different from symbol shapes appearing on paper charts. It was therefore of interest to determine whether there were key features that pilots considered to define a symbol type, regardless of display format.

There was also interest in guidance on how symbols could be modified so that one symbol may be encoded with multiple features that provide information about the symbol. FAA NACO had proposed a set of rules to ICAO that recommended a consistent way to encode symbols [1]. The rules are illustrated in Table 1 below.

Symbol shape: Navigation equipment	Fill	
	Unfilled → On-request	Filled → Compulsory
Triangle → Ground-based	△	▲
Waypoint → RNAV	✧	✦

Table 1. Symbol-Feature Rule Summary.

As shown in Table 1, the rules propose the following modifications to a symbol:

- symbol fill distinguishes between compulsory (filled) and on-request (unfilled) reporting points,
- the presence or absence of a circle surrounding the symbol distinguishes between fly-by (no circle) and fly-over (circle) waypoints, and
- symbol shape distinguishes between ground-based and GPS/Area Navigation (RNAV) waypoints.

If adopted by ICAO, the symbol-feature rules could be incorporated into international recommendations. As a result, a priority for the FAA was to determine if the recommendations were appropriate by having research focus on whether pilots could learn and apply them.

Two experiments were designed and implemented. Experiment 1 addressed the issue of symbol stereotypes, with the goal of determining the acceptable variations in a symbol's design. Experiment 2 addressed the proposed symbol-feature rules to determine the extent to which users can interpret symbol meaning as symbols become more complex. This report describes and documents the results of these studies. Suggestions to manufacturers and researchers who are interested in evaluating symbology are provided at the end of this report.

2 EXPERIMENT 1: SYMBOL STEREOTYPES

Symbols are designed to have a basic shape or characteristic that can be recognized by users. A symbol is distinctive if it is easy to discriminate from other symbols, even if it differs from other symbols by only one feature. The distinctiveness of a symbol may be measured by assessing the degree to which it can be identified within the chart and outside the context of a chart. If the symbol is identifiable *only* within the context of the chart, then the pilot may be relying on contextual clues (e.g., the location of the symbol on a chart) to determine what the symbol represents, if the meaning is not obvious. By removing all contextual clues, it is possible to discern the meaning conveyed by the symbol itself.

Consistency in symbol design across chart providers will facilitate recognition and interpretation of the symbols. However, the design of symbols has historically resulted in differences from one manufacturer to another and in differences between symbols shown on electronic displays and those used on paper charts. An example, comparing the representation of the VORTAC symbol, is shown in Table 2 below.

	USA Symbol (NACO)	Jeppesen Symbol	SAE ARP 5289
VORTAC	⬡	⊙	⊙

Table 2. Variations in the VORTAC symbol.

Note: Jeppesen symbology, Copyright 2004 Jeppesen Sanderson, Inc.

The USA symbol is used by the FAA NACO on paper charts, the Jeppesen symbol is a prototype for use on electronic charts, and the SAE symbol is recommended for use on electronic moving map displays. The USA symbol and Jeppesen symbol are different but both share commonalities with the symbol recommended in SAE ARP 5289. However, the symbols are different enough that pilots may not realize that they all represent a VORTAC.

Inconsistency in symbols across chart providers can produce cases where a symbol shape used by one provider is confusable with a symbol shape from a different provider. This confusion may result when symbols from two chart providers are visually similar but used in different ways. For example, in 1999, the FAA identified the potential for confusion due to the similarity between the USA representation for a fly-by waypoint and the previous ICAO representation for a fly-over waypoint. The symbols that were in use then are shown in Table 3.

	USA Symbols	Previous ICAO Symbols
Fly-By Waypoint	✦	✧
Fly-Over Waypoint	✦ in circle	✦

Table 3. USA and ICAO fly-by and fly-over waypoint symbols in 1999. The ICAO symbol was changed in 2000.

The USA fly-by waypoint symbol and the previously recommended ICAO fly-over waypoint symbol were both four-pointed stars with filled endpoints. The operational meanings for a fly-by waypoint versus a fly-over waypoint are significantly different, however. A *fly-by* waypoint allows the pilot to anticipate a turn to avoid overshooting the next flight segment. A *fly-over* waypoint is used when the aircraft must fly over the point prior to initiating a turn. Fly-over waypoints are usually designated because of an obstacle clearance requirement. Consequently, if these symbols are misinterpreted by a pilot, the resulting flight path deviation could have safety implications. This inconsistent implementation of waypoint symbology was addressed by ICAO and resolved in 2000 by the addition of a circle to their recommended fly-over waypoint symbol (see ICAO *Annex 4*, 10th edition [2]).

Symbol recognition is determined by the symbol's key defining features, i.e., features unique to that symbol. For example, if the key feature for the VORTAC symbol shown in Table 2 is its overall shape, then pilots who are familiar with USA symbol may not recognize the VORTAC symbol recommended in SAE ARP 5289. The goal of the current experiment was to determine what key features are necessary to recognize eight navigation symbols: DME, fix, NDB, TACAN, VOR, VORDME, VORTAC, and waypoint. Note that while fixes and waypoints are both types of intersections, a *fix* is defined by the intersection of pathways referenced to ground-based navigation aids whereas a waypoint is defined by latitude and longitude coordinates.

2.1 Method
2.2 Participants

Seventy-three active instrument-rated pilots participated in the experiment. Pilots were recruited directly from airlines, the Air Line Pilots Association (ALPA), military, and local flying clubs. Participants had a range of flying experience; 41 were air transport pilots, 14 were military pilots, 12 were GA pilots, and 6 were pilots working at the FAA who had a mix of air transport and/or military flying experience. Since the symbols pilots are familiar with will vary depending on the charts they use, participants were asked to indicate the primary chart provider for the charts they used most often. Twenty-seven pilots considered themselves USA NACO chart users, and 46 considered themselves Jeppesen chart users. However, pilots may sometimes use charts from other providers. Nine of the USA NACO chart users indicated that they had experience with Jeppesen charts. Similarly, 14 of the Jeppesen chart users had experience with the USA NACO charts. However, many of these pilots indicated that their use of these "secondary" charts, i.e., charts other than those manufactured by their primary chart provider, was infrequent.

2.2.1 Symbols

The experiment addressed the eight key navigation symbol types (DME, fix, NDB, TACAN, VOR, VORDME, VORTAC, and waypoint). The symbol shapes were collected from five aviation display manufacturers and published documents (FAA NACO *Aeronautical Chart User's Guide* [4], ICAO *Annex 4* [2], and SAE ARP 5289 [6]). Permission was obtained from display manufacturers and chart providers to use their symbols in the study.

Most of the symbols were presented in the color provided by the manufacturer. Because most moving map and navigation displays today are drawn on a black background, most of the symbols collected for the experiment were already drawn on a black background. A few symbols, however, were drawn in black on a white background. For consistency, these symbols were modified for the experiment and presented as white symbols on a black background. No other modifications to colors of symbols were made.

Foils, i.e., "fake" symbols that are not currently in use, were also presented. Responses to the foils were used as an indicator as to whether or not participants discriminated between shapes. For example, some pilots may not associate a definitive shape for a symbol type, but rather expect and accept variation in the presentation of symbols. If this were the case, then pilots would judge the foils to be acceptable.

2.2.2 Tasks

The experiment consisted of eight blocks, with each block addressing one of the symbol types. For each block, participants were shown a series of test symbol shapes and instructed to indicate whether the test symbol was representative of the symbol type. Pilots completed two tasks: *symbol recognition* and *symbol recall*.

2.2.2.1 Symbol Recognition

The first task required *symbol recognition*. Participants were shown test symbol shapes and asked to indicate whether they would consider it to be representative of the symbol type. Two versions of this task were developed: an electronic version and a paper questionnaire.

In the electronic version of the task, participants were shown the test symbol shapes one at a time *without context* on a laptop computer. Participants were asked the following:

> *Based on your knowledge of charts and navigation displays, decide whether the symbol would represent a **symbol type** or not if you saw it on a chart or navigation display.*

A trial was the presentation of a test symbol shape. Test symbol shapes were shown in isolation on a black background. Each trial was preceded by a black screen with a white crosshair in the center. This was presented for approximately 250 ms. Then, the crosshair was removed and the test symbol shape appeared, centered on the display. Participants gave a yes/no response to the test symbol using the arrow keys on the keyboard. The arrow keys were labeled "yes" or "no" to prevent confusion. Participants then provided a rating of confidence in their response. Confidence was measured on a 7-point scale, with 1 = Not confident and 7 = Very confident. Participants entered their confidence rating using the number keys on the keyboard.

For each symbol type, participants were shown 24 different test symbol shapes. Since the size at which the symbol is shown on a navigation or moving map display may vary, the symbol shapes shown in the electronic version of the symbol recognition task were presented in two sizes: small, approximately 0.125 in (0.318 mm), and large, approximately 0.25 in (0.635 mm). There were a total of 48 experimental trials.

In the paper questionnaire, participants were asked to answer the same question posed in the electronic version:

> *Based on your knowledge of charts and navigation displays, decide whether the symbol would represent a **symbol type** or not if you saw it on a chart or navigation display.*

The paper questionnaire included a set of 24 test symbol shapes for each symbol type. The test symbol shapes were identical to those presented in the electronic version of the task. However, the symbol size was not manipulated. The symbol size shown in the paper questionnaire was representative of the actual size with which the symbol would be displayed. Participants were asked to cross out the test symbol shapes that they did not consider to be representative of the symbol type.

2.2.2.2 Symbol Recall

The second task required *symbol recall*. Participants were asked to draw the symbol shape(s) that they considered to be most representative of the symbol type and state the rule they used in classifying the symbols in the electronic and/or paper symbol recognition task. The symbol recall task was presented on paper only.

2.2.3 Procedure

The experiment took approximately one hour, during which participants completed a background questionnaire (see Appendix A: Background Questionnaire), the electronic and paper versions of the symbol recognition task, and the symbol recall task. The electronic version of the symbol recognition task was available on a laptop only and administered by an experimenter. An example of the paper symbol recognition and symbol recall tasks are available in Appendix B: Symbol Stereotypes Questionnaire.

Participants completed all tasks for one symbol type before moving on to the next symbol type. The order in which the symbol types were presented was counterbalanced between subjects.

Because the electronic version of the symbol recognition task could not be self-administered, participation in the electronic version of the task was limited to pilots local to the Boston area or pilots flying through the Boston area. Additionally, there was one data collection trip to the FAA in Washington D.C. In total, 28 pilots recruited from local flying clubs, the FAA, and ALPA completed both the electronic and paper versions of the symbol recognition task and the symbol recall task. Participants who completed both tasks were given a $30 gift certificate to Sporty's Pilot Shop to thank them for their time and participation.

Initial analyses showed a high correlation in participants' responses to the electronic and paper symbol recognition tasks. As a result, in order to increase the number of pilots participating in the study, the paper questionnaire, consisting of the paper version of the symbol recognition task and symbol recall task, was distributed to 200 additional pilots recruited through airlines, ALPA, and the USA Air Force. Completing the paper version only took approximately 20 minutes. Of the 200 questionnaires distributed, 45 were returned (a 22.5% response rate).

2.3 Data

The *symbol recognition* task data were used to calculate the frequency with which a test symbol was considered representative of a symbol type. Since responses to the electronic and paper symbol recognition tasks were highly correlated, the data were combined. The frequency data was analyzed with a chi-square test to determine whether the frequency of "yes" responses was significantly higher than what would be expected from a random split. Based on the results of the analysis, the test symbols were categorized into three groups:

- *Representative symbols*: test symbols that were considered by pilots to be representative of the symbol type ("yes")

- *Mixed results*: mix of "yes"/ "no". That is, the test symbols did not receive enough "yes" responses to be considered to be representative of the symbol type but also did not receive enough "no" responses to be considered not representative of the symbol type.

- *Not representative*: test symbols that were *not* considered to be representative of the symbol type ("no")

Note that no comparison was conducted to determine a "single most stereotypical" symbol.

Shapes drawn by pilots in the *symbol recall* task were collected, categorized by shape, and counted. Since pilots sometimes drew more than one "representative" symbol, the total number of symbol shapes drawn may be greater than the number of pilots who participated in the study. As part of the symbol recall task, pilots were also asked to write the rule(s) they used to classify the test symbol shapes for the symbol recognition task. These rules often described what a symbol type looked like and are best described by the shapes pilots drew in the symbol recall task. As a result, the written rules will not be presented in detail here.

2.4 Results

The results indicate which symbol shapes were considered to be representative of the symbol type. Results for the eight navigation symbols are presented below in the next 8 sections (2.4.1 through 2.4.8). In each section, a pair of tables is presented. The first table in the pair shows how pilots classified the test symbol shapes in the symbol recognition task. *All the test symbol shapes shown in these tables are real symbols that are currently **in use**, unless otherwise indicated.* The second table in the pair shows the shapes drawn by pilots for each symbol type in the symbol recall task. The total number of times a shape was drawn is indicated. Interestingly, pilot experience and chart training did not influence which symbols were easiest to recognize and use. The representative shape identified did not differ as a function of experience (e.g., air transport, general aviation, or military) or chart provider (e.g., Jeppesen, NACO).

2.4.1 DME

Table 4 shows the classification of test symbol shapes for the DME symbol from the *symbol recognition* task. No representative shape for the DME symbol was identified in the aggregate results. Pilots commented that they typically see a DME in combination with another symbol; consequently, pilots were not familiar with the shape of a stand-alone DME.

Table 5 shows pilots' drawings of shapes they considered to be representative of a DME symbol from the recall task. The frequency with which each shape was drawn and the variety of shapes drawn confirm the results of the symbol recognition task and suggest that pilots do not have a clear stereotypical shape for a DME. The square shape, which received a mix of "yes" and "no" responses and was classified in the symbol recognition task as falling into the *mixed* category, was drawn the most frequently in the *symbol recall* task. Pilots' drawings and written rules indicate that pilots considered the DME to be a *square* (the first symbol in Table 5), a *starburst symbol* (the second symbol), or a *TACAN* symbol (the third symbol). These three shapes are in use by various manufacturers and chart providers for representing a DME. Conversely, the three symbols drawn by only one pilot each (the bottom three drawings in Table 5) either do not exist or are not used to represent a DME.

Table 4. Symbol Recognition: DME.

Note: Jeppesen symbology, Copyright 2004 Jeppesen Sanderson, Inc. Some symbols are reduced or use different colors for illustrative purposes. Specifically with reference to color, the symbols provided by Jeppesen were inverted from being black symbols on a white background to white symbols on a black background.

DME Basic Shapes Drawn	TOTAL
	33
	16
	16
	5

DME Basic Shapes Drawn	TOTAL
	3
	1
	1
	1

Table 5. Symbol Recall: DME.

2.4.2 Fix

Table 6 shows the classification of test symbol shapes for a fix in the symbol recognition task. Table 7 shows pilots' drawings of shapes they considered to be representative of a fix from the symbol recall task.

The results of both the symbol recognition and symbol recall tasks show that pilots consider a triangle shape to be representative of a fix. As shown in Table 6, pilots' classified the test symbol shapes despite the variations in the size, fill, or color of the symbol shape. Note that some manufactures and chart providers (and even some pilots) do not distinguish between symbols for fixes and waypoints. In fact, the first three symbols drawn in Table 7 are all in use to represent a fix. Conversely, the bottom two symbols in Table 7, drawn by only one pilot each, either do not exist or are not used to represent a DME.

Symbol Type	Representative	Mixed	Not Representative
Fix	△ ▲ △ ▲ △ △ △ △	△ Ⓐ Ⓐ Ⓐ	✦ ✦ ◇ ◇ ◉ ✕

Table 6. Symbol Recognition: Fix.

Note: Jeppesen symbology, Copyright 2004 Jeppesen Sanderson, Inc. Some symbols are reduced or use different colors for illustrative purposes. Specifically with reference to color, the symbols provided by Jeppesen were inverted from being black symbols on a white background to white symbols on a black background.

FIX Basic Shapes Drawn	Total
△	83
✦	13
✕	7
☆	1
◈	1

Table 7. Symbol Recall: Fix.

9

2.4.3 NDB

Table 8 shows the how test symbol shapes representing the NDB were classified in the symbol recognition task. Table 9 shows pilots' drawings of shapes they considered to be representative of a NDB in the symbol recall task.

As shown in the tables below, pilots considered the "representative" NDB shape to be *an array of small dots* with a circle in the center. The "representative" NDB shape did not differ based on whether the circle in the center was filled or unfilled, the size of the circle in the center, or whether or not the symbol was surrounded by a circle. As the array of dots became less distinctive (e.g., see the top symbol in the Mixed category column), pilots were not sure if the symbol was an NDB or not. Test symbol shapes shown in Table 8 that were considered to be not representative of an NDB were probably a result of the fact that the array of dots was not present.

Table 8. Symbol Recognition: NDB.

Note: Jeppesen symbology, Copyright 2004 Jeppesen Sanderson, Inc. Some symbols are reduced or use different colors for illustrative purposes. Specifically with reference to color, the symbols provided by Jeppesen were inverted from being black symbols on a white background to white symbols on a black background.

NDB Basic Shapes Drawn	Total
(array of dots with circle)	70
(circle)	7
(triangle)	5
(arrows)	1

Table 9. Symbol Recall: NDB.

2.4.4 TACAN

The TACAN symbol is used primarily by the military, but a representative shape was identified even though many non-military pilots commented that they did not use the symbol. Table 10 shows how the test symbol shapes representing the TACAN were classified in the symbol recognition task. Table 11 shows pilots' drawings of shapes they considered to be representative of a TACAN in the symbol recall task.

Pilots' rules for the TACAN described the "representative" shape as a *Y-shaped* symbol or *a three pronged object with flattened points to prongs and curved webbing between prong points*. This shape is in the "Representative" column in Table 10 and was drawn most frequently in the symbol recall task, shown in Table 11. Note that the results of the symbol recognition task in Table 10 show that pilots identified a representative shape despite the variations in the size, color, and fill of the test symbol shapes.

Symbol Type	Representative	Mixed	Not Representative
TACAN			

Table 10. Symbol Recognition: TACAN.

Note: Jeppesen symbology, Copyright 2004 Jeppesen Sanderson, Inc. Some symbols are reduced or use different colors for illustrative purposes. Specifically with reference to color, the symbols provided by Jeppesen were inverted from being black symbols on a white background to white symbols on a black background.

TACAN Basic Shapes Drawn	Total
	55
	13
	7

Table 11. Symbol Recall: TACAN.

2.4.5 VOR

Table 12 shows the classification of test symbol shapes for a VOR in the symbol recognition task. Table 13 shows pilots' drawings of what shapes they considered to be representative of a VOR from the symbol recall task. The tables show that pilots considered the "representative" shape of a VOR to be a hexagon, regardless of variations in the size, color, and fill of the test symbol shapes. Note in Table 12, one foil, created with two concentric hexagons, received a mix of "yes" and "no" responses. It is likely that some pilots considered the foil to be a VOR because the overall shape of the foil matched the representative shape.

Symbol Type	Representative	Mixed	Not Representative
VOR	(hexagons of various sizes and fills)	(hexagon/compass rose symbols, including foil)	(circles, crosses, compass rose symbols)

Table 12. Symbol Recognition: VOR.

Note: Jeppesen symbology, Copyright 2004 Jeppesen Sanderson, Inc. Some symbols are reduced or use different colors for illustrative purposes. Specifically with reference to color, the symbols provided by Jeppesen were inverted from being black symbols on a white background to white symbols on a black background.

VOR Basic Shapes Drawn	Total
(hexagon)	53
(compass rose)	22
(hexagon variant)	3
(Y-shape)	1
(irregular hexagon)	1

Table 13. Symbol Recall: VOR.

2.4.6 VORDME

Table 14 shows the how the test symbol shapes representing the VORDME were classified in the symbol recognition task. Table 15 shows pilots' drawings of shapes they considered to be representative of a VORDME in the symbol recall task.

Pilots' rules described the representative shape for a VORDME as a hexagon surrounded by a square. As Table 14 shows, pilots identified a representative shape despite the variations in the size and color of the shape. The fill of the center and the presence of a circle surrounding the symbol did introduce some uncertainty in the classification, as shown by the first two symbols in the "Mixed" column; pilots were not sure if these filled symbols were VORDMEs. Note that in Table 15, eight pilots drew the individual components that make up a VORDME; five pilots drew the symbol shape for a stand-alone VOR only and three drew the shape for a stand-alone DME.

Symbol Type	Representative	Mixed	Not Representative
VORDME			

Table 14. Symbol Recognition: VORDME.

Note: Jeppesen symbology, Copyright 2004 Jeppesen Sanderson, Inc. Some symbols are reduced or use different colors for illustrative purposes. Specifically with reference to color, the symbols provided by Jeppesen were inverted from being black symbols on a white background to white symbols on a black background.

VORDME Basic Shapes Drawn	Total
	32
	20
	15
	7
	5
	3
	1

Table 15. Symbol Recall: VORDME.

2.4.7 VORTAC

Pilots' rules described a VORTAC as *a hexagonal shape with three of the tips blocked*. The rule is shown in the results of the symbol recognition task in Table 16 and symbol recall task in Table 17. Table 16 shows that pilots identified a representative shape in the symbol recognition task regardless of the variations in the size, fill, and color with which the test symbol shapes were presented. Note that a foil was categorized as being representative of a real VORTAC symbol (see the test symbol in the bottom row of Table 16). This foil had the same overall shape as a representative VORTAC symbol but was rotated 180°. This result suggests that pilots do not consider the orientation of the symbol as a critical aspect of the coding. That is, a symbol can still be a VORTAC if it looks the same but is rotated. A second foil received a mix of "yes" and "no" responses; this foil, shown in the bottom row of the "Mixed" column in Table 16, is triangular with rounded, filled endpoints, but the overall shape and fill pattern is similar enough to that of the representative symbols that the foil was considered to be representative of a VORTAC symbol 47% of the time.

Table 16. Symbol Recognition: VORTAC.

Note: Jeppesen symbology, Copyright 2004 Jeppesen Sanderson, Inc. Some symbols are reduced or use different colors for illustrative purposes. Specifically with reference to color, the symbols provided by Jeppesen were inverted from being black symbols on a white background to white symbols on a black background.

VORTAC Basic Shapes Drawn	TOTAL
	64
	9
	3

VORTAC Basic Shapes Drawn	TOTAL
	1
	1

Table 17. Symbol Recall: VORTAC.

2.4.8 Waypoint

Table 18 shows the how the test symbol shapes representing the waypoint were classified in the symbol recognition task; Table 19 shows pilots' drawings of shapes they considered to be representative of a waypoint in the symbol recall task.

The classification of symbols (shown in Table 18), drawings (shown in Table 19), and written rules all indicate that pilots' consider the representative shape for a waypoint to be a four-pointed star. As Table 18 shows, the representative shape was identified despite variations in the size, fill, color, or presence of a circle surrounding the symbol.

Symbol Type	Representative	Mixed	Not Representative
Waypoint	(various four-pointed star symbols)		(triangle, diamond, diamond)

Table 18. Symbol Recognition: Waypoint.

Note: Jeppesen symbology, Copyright 2004 Jeppesen Sanderson, Inc. Some symbols are reduced or use different colors for illustrative purposes. Specifically with reference to color, the symbols provided by Jeppesen were inverted from being black symbols on a white background to white symbols on a black background.

Waypoint Basic Shapes Drawn	Total
(four-pointed star)	77
(triangle)	5
(shaded diamond)	1

Table 19. Symbol Recall: Waypoint.

2.4.9 Summary

Symbol shape was key factor for classifying symbols in the *symbol recognition* task. The representative shapes identified from the electronic version of the symbol recognition task, the paper version of the symbol recognition task, and the symbol recall task were identical – a finding which strongly supports the idea that pilots have stereotypes for symbols and that those stereotypes were identified by the current study. Table 20 summarizes the representative shape identified for each symbol type.

Symbol Type	"Representative Shape"
DME	None Identified
Fix	△
NDB	⊛
TACAN	⬡
VOR	⬡
VORDME	▢
VORTAC	⬡
Waypoint	✧

Table 20. Symbol Stereotypes: Representative shape.

Note: Jeppesen symbology, Copyright 2004 Jeppesen Sanderson, Inc. Some symbols are reduced or use different colors for illustrative purposes. Specifically with reference to color, the symbols provided by Jeppesen were inverted from being black symbols on a white background to white symbols on a black background.

As Table 20 shows, a representative symbol shape was identified for seven of the eight symbols. No representative shape was identified for the DME in the aggregate results. Pilots' comments indicated that DMEs are typically drawn in conjunction with another symbol, so they were less familiar with the shape of a stand-alone DME.

Shape appears to be the important factor in determining whether a symbol shape is representative of a symbol type. The results indicated that the size, color, and orientation were not critical factors in determining what the symbol was. Representative shapes were identified despite the variations in size, color, and orientation with which the test symbol shapes were presented. Symbol fill generally did not influence pilots' ratings as to whether a symbol shape was representative of the symbol type, but circles surrounding symbols created some uncertainty. The presence of a circle surrounding a symbol is a convention used by some chart providers to distinguish a fly-over symbol from a fly-by symbol. (This coding method will be discussed in more detail in the next experiment). However, this convention is not yet in widespread use[1]. Consequently, pilots may not have known whether the circle was a feature of the symbol or created a different symbol entirely. Therefore, while the results show that symbols surrounded by a circle tended to fall in the "mixed" category, the results should not be interpreted to speak to the usability of the circle rule.

The test symbol shapes presented in this study consisted of symbols used on paper charts, electronic charts, and electronic navigation displays. It is interesting to note that the representative symbols

---

[1] Sixteen of the pilots who participated in Experiment 1 were previously exposed to the circle feature-rule, but this did not appear to influence the results.

identified (and shown in Table 20) are commonly used on FMS and moving map displays, with the exception of the NDB symbol, and NACO paper charts. Thus, the representative shapes identified are most likely due to pilots' familiarity with the symbols shown on electronic displays, regardless of their chart provider.

Thus, the results highlight the importance of consistency in symbol design not only across chart providers but also across display mediums. This issue of consistency will become more important as electronic charts replace existing paper charts in the future. Consistency is also important in symbol design as the design increases in complexity. While high levels of detail may convey more information about a symbol, the additional complexity may inhibit the usability of the symbol. Experiment 2 addresses these issues and the extent to which users can interpret symbol meaning as symbols become more complex.

3 EXPERIMENT 2: SYMBOL-FEATURE RULES

Symbols may be designed so that one symbol has multiple features that convey information. For example, a single symbol can be used to convey not only what the symbol represents (e.g., a VOR) but also if it is a compulsory reporting point or not (e.g., a filled symbol is compulsory and an unfilled symbol is on-request). Recently, the USA submitted a proposal to ICAO to establish a set of symbol-feature rules that define a consistent method for coding symbols to distinguish between compulsory versus on-request reporting, fly-over versus fly-by requirements, and ground-based points versus RNAV points for navigation purposes (for more information, see ICAO Aeronautical Information and Charts Study Group Meeting Working paper, Concept for RNAV/Ground-Based Charting Symbol Consistency and Hierarchy, [1]). The proposed symbol-feature rules are shown in Table 21.

Symbol shape = Navigation equipment		Fly-By → no circle		Fly-Over → circle	
		Unfilled → On Request	Filled → Compulsory	Unfilled → On Request	Filled → Compulsory
Ground-based	DME	□	■	⊡	⊡
	Fix	△	▲	◬	◬
	NDB	⊙	⊙	⊙	⊙
	TACAN	⬠	⬟	⬠	⬟
	VOR	⬡	⬢	⬡	⬢
	VORDME	▢	▣	▢	▣
	VORTAC	⬡	⬢	⬡	⬢
RNAV	Waypoint	✦	✦	✦	✦

Table 21. Symbols defined by combining shape and fill.

As Table 21 shows, the *symbol shape* designates the navigation-equipment requirement. The DME, fix, NDB, TACAN, VOR, VORDME, and VORTAC define ground-based navigation aids, and the four-pointed waypoint star identifies the location of RNAV waypoints, defined by latitude/longitude coordinates. The shape feature is combined with fill to indicate whether the point is an on-request or compulsory reporting point. For example, an unfilled triangle represents on-request reporting at a ground-based point, and a filled triangle represents compulsory reporting at a ground-based point. A third feature is the presence or absence of a circle that surrounds the symbol. The circle differentiates between a fly-over requirement (circle) and a fly-by requirement (no circle).

When expanding the definitions of current symbols and/or integrating symbols, the design of the symbol needs to be evaluated to determine whether the meaning conveyed by the combined features is clear. The amount of instruction that pilots are given about how the symbol rules are applied may influence their ability to learn the rules. The more information pilots are given, the easier it will be to learn the coding scheme. While this information will typically be available in the chart legends, its effect is not clear because many pilots do not refer to their legends while using their charts.

It will also be important to consider whether the application of the symbol-feature rule is consistent with previous training. That is, the new definition of a symbol should be consistent with how the symbol is currently used. One potential issue in the application of the USA-proposed symbol-feature rules is in the definition of the triangle symbol. Currently, the triangle symbol is used to represent a fix, which can be navigated to without RNAV equipment. The USA-proposed symbol-feature rules change the meaning of the triangle, however, so that the triangle symbol represents only a ground-based reporting point. Consequently, it will be important to determine what the implications are, if any, for changing the meaning of the triangle symbol.

The goal of this experiment is to determine the effects of complexity in symbol design and the extent to which users can interpret symbol meaning as symbols become more complex. The amount of information pilots are given about the rules was varied to determine the effect of instruction on the learnability of the rules.

3.1 Method
3.1.1 Participants

Pilots were recruited from local flying clubs and the Air Line Pilots Association (ALPA). Twenty-one current instrument-rated pilots participated in the experiment. Thirteen were GA pilots; eight were air transport pilots. Participants were given a $30 gift certificate to Sporty's Pilot Shop to thank them for their time and participation.

3.1.2 Symbols

Eight symbol types (DME, fix, NDB, TACAN, VOR, VORDME, VORTAC, and waypoint) served as *base symbols*. Two versions of the base symbols were tested: those recommended in NACO's *Aeronautical Chart User's Guide, 6th Edition* [4] and those recommended in ICAO *Aeronautical Charts, Annex 4* [2] for paper charts. Note that the recommended shape for a fix symbol is the same in both documents.

The base symbols were modified to create four test symbols. The test symbols consisted of:
- a fly-by, on-request reporting point;
- a fly-by, compulsory reporting point;
- a fly-over, on-request reporting point;
- and a fly-over, compulsory reporting point.

ICAO *Annex 4* [2] includes symbol recommendations for electronic displays as well. The recommended electronic symbols differ slightly in some cases from the recommendations for paper charts. Four electronic symbols, representing the NDB, VOR, VORDME, and VORTAC, were included in the current study. The recommended electronic symbol shape for these symbol types is shown in the second column of Table 22 below. The recommended electronic symbol shape was modified to create test symbol shapes based on the proposed symbol-feature rules. Note that these test symbol shapes are *not* included in ICAO *Annex 4*. The VOR and VORDME symbols were modified to create a fly-by, on-request symbol and a fly-over, on-request symbol. The VORTAC and NDB symbols were modified to create a fly-by, on-request symbol; a fly-over, on request symbol; and a fly-over, compulsory symbol.

	ICAO Recommended Electronic Symbol	Test Symbol Shapes		
		Fly-By On-request	Fly-Over On-Request	Fly-Over Compulsory
NDB	⊙	⊙	⊙	⊙
VOR	⊕	⊕	⊕	
VORDME	⊟	⊟	⊟	
VORTAC	⊕	⊕	⊕	⊕

Table 22. ICAO electronic symbols and their tested versions.

To determine whether or not participants could generalize and apply the symbol-feature rules, test symbol shapes that are not currently in use on navigation displays, i.e., *foils* or "fake" symbols, were created. Each base foil shape was modified to create four test symbols. An example is shown below in Table 23.

	Fly-By		Fly-Over	
	On-request	Compulsory	On-Request	Compulsory
Foil	△	▲	⊕	⊕

Table 23. Foil Example.

In all, 90 unique test symbols were presented. The test symbols consisted of 60 *paper-based* symbols, 10 *electronic* symbols, and 20 *foils*. Each test symbol was then drawn in two sizes: a small size which was approximately 0.125" (0.318 mm, similar to the size of symbols on moving map displays), and a large size which was approximately 0.25" (0.635 mm, similar to the size of symbols on FMS displays).

3.1.3 Task

Each of the three symbol-feature rules was addressed individually. Pilots were shown test symbols on a laptop computer and asked to answer one of the following questions:

- Is the symbol fly-over or fly-by?
- Is the symbol a compulsory or on-request reporting point?
- Is the symbol a ground-based or RNAV fix?

The order in which the questions were presented was counterbalanced across participants.

For each rule, participants were shown an introductory slide that defined the feature being addressed and provided with a legend depicting an application of the rule. Examples are shown in Figure 1 and Figure 2.

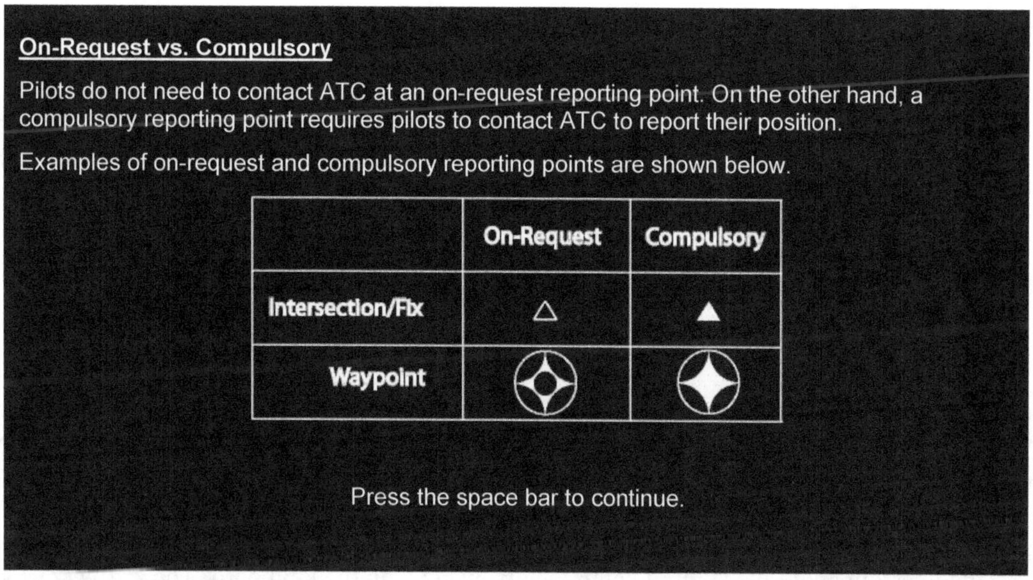

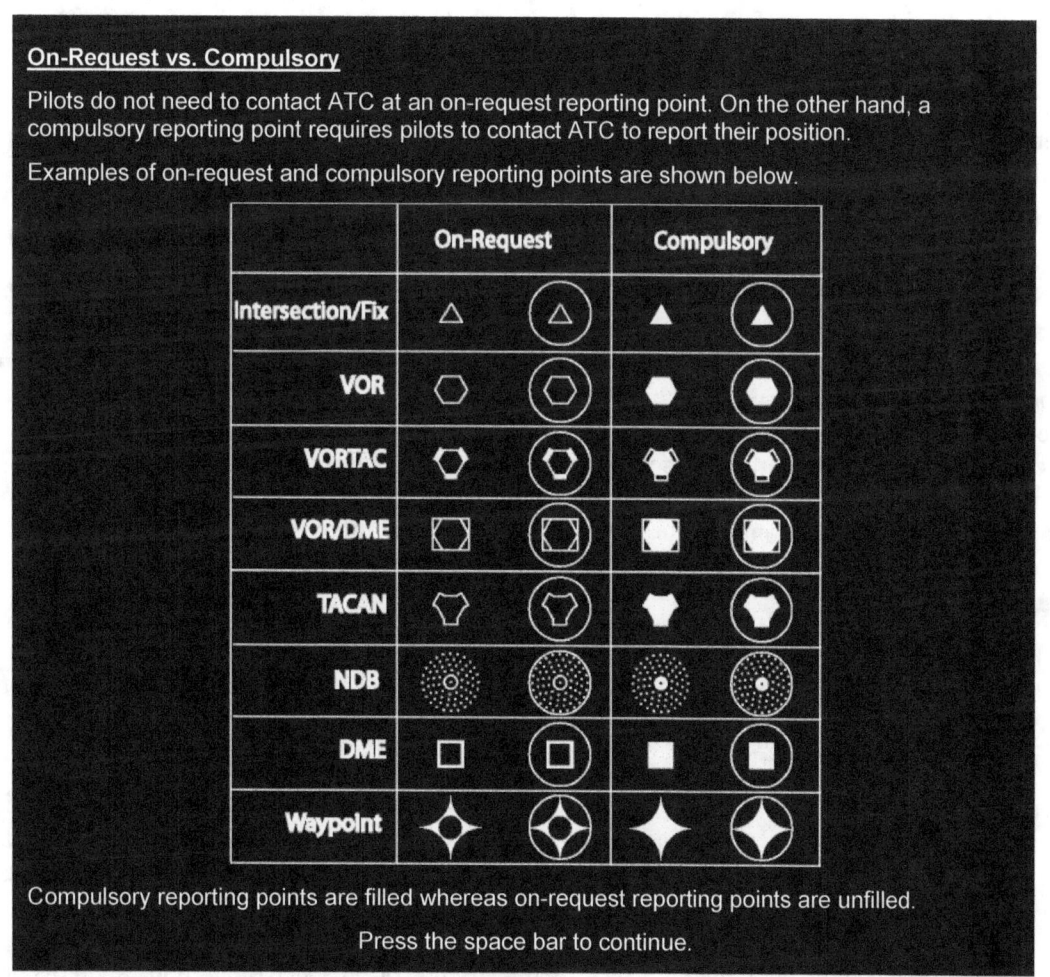

Figure 1. Introductory slide: Minimal legend depicting the on-request vs. compulsory reporting point rule.

Figure 2. Introductory slide: Detailed legend depicting the on-request vs. compulsory reporting point rule.

Since the amount and type of instruction that participants are given about how the symbol rules are applied may influence their ability to learn the rules, the legend was varied at two levels of detail: a *minimal* legend, shown in Figure 1, that showed two examples of the rule without explicit instruction, and a *detailed* legend, shown in Figure 2, that provided an explicit, detailed description of the rules and examples of the rule applied to the eight base symbols. The level of detail provided in the symbol legend was varied between participants.

3.1.4 Procedure

The experiment took approximately 1 hour to complete, during which time participants completed a background questionnaire (see Appendix A: Background Questionnaire) and the tasks for each of the three rules.

The experiment was a mixed design; symbol type and symbol size were manipulated within subjects and symbol legend was varied between subjects. Test symbol shapes were presented in isolation on a black background. A trial, as defined here, was the presentation of a test symbol shape. Each trial was preceded by a black screen with a white crosshair in the center, shown for approximately 250 ms. Then, the crosshair was removed and the stimulus appeared centered on the display. Participants then recorded their response to the question about the symbol using the arrow keys on the keyboard. The arrow keys were highlighted with stickers and the mapping of the responses (e.g., left arrow = fly-by, right arrow = fly-over) was indicated on an index card placed below the keys. Once participants gave their response, they were asked to provide a rating of confidence on a 1-7 scale (1 = not confident, 7 = very confident). Pilots entered their confidence rating using the number keys on the keyboard.

For each rule, participants were shown 18 practice trials before the actual data collection trials. Upon completion of the practice trials, participants were shown two experimental blocks containing 90 trials each. All participants saw all the test symbols.

3.2 Data

The independent variables were the legend detail (minimal, detailed), symbol type (DME, NDB, fix, TACAN, VOR, VORDME, VORTAC, and waypoint), and symbol size (small, large). The dependent variables were the classification accuracy, the response time, and participants' rating of confidence in their response. For the purposes of the data analysis, a *confidence score* was created by weighting participants' rating of confidence with their accuracy for that trial. For correct responses, a multiple of +1 was used; for incorrect responses, a multiple of –1 was used. Thus, participants who were accurate had a higher confidence score than participants who were inaccurate. Additionally, participants who knew that they did not know the answer or knew that they had responded incorrectly, i.e., participants who were not confident in their response and inaccurate, had a higher confidence score than participants who did not know that they did not know the answer or were confident but inaccurate.

3.3 Results

The following sections address the symbol-feature rules to determine how easy the rules are to understand, learn, apply, and remember. Section 3.3.1 examines the use of the fly-by vs. fly-over rule; Section 3.3.2 addresses the use of the compulsory vs. on-request rule; and Section 3.3.3 discusses the use of the ground-based vs. RNAV rule.

The mean accuracy, response time, and confidence scores were calculated for each paper, electronic, and foil symbol and used to evaluate the rules and identify if any symbols were problematic. The results showed no effects attributable to symbol size, so the data was collapsed across that factor.

3.3.1 Fly-By vs. Fly-Over

Fly-by and fly-over waypoints are distinguished by the absence or presence of a circle: symbols drawn without a circle are fly-by waypoints; symbols surrounded by a circle are fly-over waypoints. Table 24 presents the mean accuracy, response time, and confidence for pilots' fly-by/fly-over classification of the paper-based symbols, electronic symbols, and foils.

Symbol Type	Symbol Shapes	Fly-By vs. Fly-Over		
		Accuracy	Response Time	Confidence Score
DME		97%	1.34s	6.13
NDB		86%	1.89s	4.54
TACAN		97%	1.35s	6.16
VOR		97%	1.33s	6.04
VORDME		93%	1.55s	5.52
VORTAC		98%	1.21s	6.19
Fix		99%	1.34s	6.32
Waypoint		98%	1.55s	6.22
Foil		97%	1.46s	6.00
Electronic		74%	1.91s	2.84

Table 24. Symbol-Feature Rules: Fly-By vs. Fly-Over. The NDB symbol was most problematic when applying the fly-by/fly-over symbol-feature rule to the paper symbols.

The analysis conducted examined (1) how intuitive the rules were to apply to paper symbols and (2) how well the rules were generalized to foils and electronic symbols, based on three dependent variables: accuracy, response time, and confidence.

3.3.1.1 Fly-By vs. Fly-Over: Paper Symbols

An 8 (symbol type) x 2 (fly-by vs. fly-over) x 2 (compulsory vs. on-request) x 2 (legend detail: minimal vs. detailed) repeated measures Analysis of Variance (ANOVA) was conducted on the data for the eight paper navigation symbols (outlined in bold in Table 24 above). The analysis revealed a significant difference in the accuracy, response time, and confidence scores to classify the symbols as a function of symbol shape [accuracy: $F(7, 119) = 5.62, p < 0.01$; response time: $F(7, 119) = 6.72, p < 0.01$; confidence score: $F(7, 119) = 8.32, p < 0.01$].

None of the higher order interactions for the accuracy and response time variables were significant ($p > 0.05$). The ANOVA did reveal, however, that the effect of symbol type on confidence score was modulated by a significant interaction between symbol type and whether the symbol was fly-by or fly-over. Paired comparisons for each symbol type showed that pilots had a higher confidence score when classifying the fix symbol as fly-over than fly-by ($p < 0.05$). There was no difference in pilots' confidence scores when classifying the other symbols ($p > 0.05$).

Of the eight symbol types, the data shows that the NDB symbol was the most problematic (the data for the NDB symbol is highlighted in Table 24). The NDB symbol was classified with less accuracy than the other symbols [this difference was significant with respect to the DME, TACAN, VOR, VORTAC, fix, and waypoint ($p < 0.05$) and marginally significant for the VORDME, $F(1, 17) = 3.15, p < 0.10$]. Additionally, the classification of the NDB as fly-by or fly-over took more time and was completed with less confidence, as defined by a lower confidence score, than the other symbols ($p < 0.05$). This difficulty classifying the NDB symbol may be attributable to three reasons. First, the existence of the circle distinguishing fly-over from fly-by may have been difficult to see given its proximity to the base symbol. Second, the global shape of the NDB symbol (i.e., circular) did not change as much as the global shape of other symbols when it was surrounded by a circle. Finally, the NDB symbol shape presented in the study was a paper-based symbol, and displaying the symbol electronically required a high resolution. Consequently, the NDB symbol may not have been depicted as clearly as the other symbols, particularly when it was presented in its *small* size, and as a result, the circle surrounding the symbol was not easy to see, i.e., not salient.

The results show that pilots were slightly more accurate (4%) in their responses and had a significantly higher confidence score when they were shown the detailed legend that explicitly described the rule and showed many examples of the rule (98% accuracy with a confidence score of 6.54) relative to the minimal legend (94% with a confidence score of 5.41) [accuracy: $F(1, 17) = 3.87, p = 0.07$; confidence score: $F(1, 17) = 5.82, p < 0.05$]. There was no difference in response time attributable to legend detail, $F(1, 17) = 0.25, p = 0.63$.

3.3.1.2 Fly-By vs. Fly-Over: Foils and Electronic Symbols

A single factor repeated measures ANOVA comparing the classification of the foils and electronic symbols relative to the paper symbols showed significant differences in accuracy [$F(2, 32) = 59.04, p < 0.01$], response time [$F(2, 32) = 9.30, p < 0.01$], and confidence score [$F(2, 32) = 89.58, p < 0.01$]. The data showed no difference in the classification of the foils versus the paper symbols. In fact, the 97% accuracy rate and high confidence show that pilots were able to apply the rules to the fake symbols.

Thus, the results suggest that the fly-by/fly-over rule was learned and could be applied, even to unfamiliar symbols. In fact, the circle/no-circle rule may be so compelling that pilots apply it, even when its use is not intended. An example is seen in the classification of the electronic symbols as fly-by vs. fly-over. Here, paired comparisons showed a difference in the classification of the electronic symbols and the classification of the paper symbols and foils ($p < 0.05$). As shown in Table 24, many of the electronic symbols are surrounded by a compass rose, i.e., a circle with the four compass points marked. Pilots may have misinterpreted the compass rose in these symbols to be a fly-over circle. The data support this hypothesis and show that pilots had a propensity to classify the electronic symbols as fly-over rather than fly-by. In fact, the accuracy rate for classifying fly-by electronic symbols was at chance level (51%), whereas the accuracy for classifying fly-over electronic symbols was 97%.

3.3.2 Compulsory vs. On-Request

Symbols representing compulsory and on-request reporting points are differentiated by *fill*: compulsory reporting points are filled, whereas on-request reporting points are unfilled. Table 25 shows the mean accuracy, response time, and confidence scores for pilots' classification of the paper-based symbols, electronic symbols, and foils as compulsory or on-request.

Symbol Type	Symbol Shapes	Compulsory vs. On-Request		
		Accuracy	Response Time	Confidence Score
DME		91%	1.68s	5.16
NDB		68%	2.55s	1.82
TACAN		90%	1.62s	5.24
VOR		93%	1.50s	5.44
VORDME		91%	1.58s	5.08
VORTAC		89%	2.02s	4.81
Fix		96%	1.72s	5.97
Waypoint		91%	2.12s	5.12
Foil		93%	1.57s	6.12
Electronic		89%	2.74s	5.65

Table 25. Symbol-Feature Rules: Compulsory vs. On-Request. The NDB symbol was most problematic when applying the compulsory/on-request symbol-feature rule to the paper symbols.

3.3.2.1   Compulsory vs. On-Request: Paper Symbols

An 8 (symbol type) x 2 (fly-by vs. fly-over) x 2 (compulsory vs. on-request) x 2 (legend detail: minimal vs. detailed) repeated measures ANOVA was conducted on the data for the paper-based symbols (outlined in bold in Table 25 above). The analysis revealed a significant difference in the accuracy, response time, and confidence scores in classifying the symbols as compulsory or on-request due to symbol shape [accuracy: $F(7, 112) = 8.78, p < 0.01$; response time: $F(7, 112) = 5.55, p < 0.01$, confidence score, $F(7, 112) = 17.53, p < 0.01$].

The analysis also showed that pilots were faster classifying symbols that represented fly-by waypoints, i.e., those not surrounded by a circle, as compulsory or on-request than those that represented fly-over waypoints, i.e., those surrounded by a circle (1.7 s versus 2.0 s respectively; $F(1, 16) = 13.83, p < 0.05$). It is possible that the presence of a circle surrounding the symbol simply added another feature to be considered when classifying the symbol, which slowed the time to respond. However, it is important to note that neither the accuracy nor confidence score data show this effect. As a result, this effect may simply be an anomaly and may not be representative of any phenomenon. None of the higher order interactions for accuracy, response time, or confidence score were significant ($p > 0.05$).

Paired comparisons showed that the NDB symbol was classified less accurately and with a lower confidence score than the other symbols. These results, which are highlighted in Table 25, were significant, $p < 0.05$. Paired comparisons on the response time data showed significant differences in the time needed to classify the NDB, VORTAC, and waypoint symbols relative to the TACAN, VOR, VORDME, and fix ($p < 0.05$); the NDB and waypoint symbols also took significantly longer to classify than the DME symbol ($p < 0.05$). There was no difference in the response time needed to classify the NDB, VORTAC, or waypoint.

The difficulty classifying the NDB symbol as compulsory or on-request may be attributable to the fact that the area that is filled or unfilled is small. Consequently, the fill of the symbol may not be salient. Additionally, because the NDB symbol is drawn with a high level of detail, it is possible that the resolution required with which the symbol was drawn was not sufficient for the fill to be salient.

The extra time needed to classify the VORTAC and waypoint relative to the other symbols may be a consequence of the partial filling of the base symbol shapes for these symbol types. That is, the "ears" of the VORTAC base symbol and the endpoints of the waypoint base symbol are filled. Consequently, participants simply could not quickly determine whether the symbol was filled or unfilled.

There was no effect of legend detail in classifying the symbols as compulsory or on-request [accuracy: $F(1, 16) = 2.62, p = 0.12$; response time: $F(1, 16) = 0.34, p = 0.57$]. The use of fill to distinguish between compulsory and on-request reporting point is already in use by some chart providers. As a result, pilots may already be familiar with the rule so that additional instruction was not necessary.

3.3.2.2 Compulsory vs. On-Request: Foils and Electronic Symbols

A single factor repeated measures ANOVA comparing the classification of the foils, electronic symbols, and paper symbols showed no overall difference in accuracy [$F(2, 32) = 0.46, p = 0.64$], response time [$F(2, 32) = 2.07, p = 0.14$], or confidence score [$F(2, 32) = 0.63, p = 0.54$]. The results suggest that the rules for distinguishing compulsory and on-request symbols could be learned and applied.

3.3.3 Ground-Based vs. RNAV

Table 26 shows the mean accuracy, response time, and confidence scores for pilots' classification of the paper-based symbols, electronic symbols, and foils as ground-based or RNAV. Ground-based and RNAV fixes are distinguished by *shape*: the DME, fix, TACAN, VOR, VORDME, and VORTAC represent ground-based fixes; the four-pointed star represents an RNAV fix.

Symbol Type	Symbol Shapes	Ground-Based vs. RNAV		
		Accuracy	Response Time	Confidence Score
DME		94%	1.34s	5.42
NDB		96%	1.27s	6.12
TACAN		99%	1.13s	6.60
VOR		100%	1.14s	6.60
VORDME		99%	1.30s	6.51
VORTAC		100%	0.98s	6.77
Fix		82%	1.73s	4.45
Waypoint		95%	1.46s	5.99
Foil		59%	1.88s	1.00
Electronic		92%	1.60s	5.37

Table 26. Symbol-Feature Rules: Ground-Based vs. RNAV. The fix symbol was most problematic when applying the ground-based/RNAV symbol-feature rule to the paper symbols.

3.3.3.1 Ground-Based vs. RNAV: Paper Symbols

A 2 (symbol legend) x 8 (symbol type) x 2 (symbol size) ANOVA revealed significant differences in accuracy, time, and confidence score in the classification of the paper symbols as ground-based or RNAV due to symbol type [accuracy: $F(7, 2510) = 21.45, p < 0.01$; response time, $F(7, 2509) = 8.50, p < 0.01$; confidence score, $F(7, 2519) = 27.43, p < 0.01$]. None of the higher order interactions for accuracy, response time, or confidence score were significant ($p > 0.05$).

Of the eight symbols, the data shows that the fix symbol was the most problematic. Results for the fix symbol are highlighted in Table 26. The fix symbol was classified less accurately and with a lower confidence score than the other paper symbols ($p < 0.05$). Paired comparisons also showed that classifying the fix symbol as ground based or RNAV took more time to classify than all the other paper symbols, except the waypoint ($p < 0.05$).

The ANOVA also revealed a benefit to the presentation of a detailed legend relative to the minimal legend. Participants who were given instructions that explicitly stated the rule and showed many examples of the rule were more accurate, $F(1, 2510) = 26.48$, $p < 0.01$, had a higher confidence score, $F(1, 2519) = 100.63$, $p < 0.01$, and faster, $F(1, 2509) = 17.49$, $p < 0.01$, in classifying the symbols.

3.3.3.2 Ground-Based vs. RNAV: Foils and Electronic Symbols

A 2 (symbol legend) x 3 (symbol class: paper, electronic, foil) x 2 (ground-based vs. RNAV) x 2 (symbol size) ANOVA was conducted to compare the classification of the paper symbols to the classification of the electronic symbols and foils. The paper symbols were classified with a mean accuracy rate of 96% in 1.27s, and a mean confidence score of 6.16. Electronic symbols were classified with a mean accuracy of 92% in 1.60s, and a mean confidence score of 5.37. Finally, foils were classified with a mean accuracy of 59% in 1.88s, and a mean confidence score of 1.00.

The data show significant differences in the classification of symbols due to the symbol class [accuracy, $F(2, 3355) = 421.32$, $p < 0.01$; response time, $F(2, 3354) = 48.36$, $p < 0.01$; confidence score, $F(2, 3355) = 554.10$, $p < 0.01$]. Not surprisingly, the paper symbols were classified with the highest accuracy, in the shortest time, and with the highest confidence scores, whereas the foils were classified with the lowest accuracy, longest time, and lowest confidence scores. The results highlight that the rules were learned and could be applied to the real symbols. While the results imply that the rules could not be applied to the foils, it is important to note that that the data for the foils *can not* be interpreted to mean that that the ground-based/RNAV rule was not intuitive since the ground-based/RNAV distinction is a function of shape. Consequently, the foils, which were different in shape than the paper symbols and did not match any of the shapes provided in the legend, were classified with an accuracy close to chance levels (50%), as would be expected.

3.3.4 Summary

The USA-proposed rules recommend a consistent way to design symbols using the following features:

- the presence or absence of a circle that surrounds the symbol to differentiate between a fly-over requirement (circle) and a fly-by requirement (no circle).

- the symbol fill to indicate whether the point is an on-request (unfilled) or compulsory reporting point (filled), and

- the symbol shape to designate the navigation-equipment requirement.

The accuracy, response time, and confidence scores for the three components of the symbol-feature rules tested in this study show that pilots were generally able to learn and apply the rules to the symbols shown. From a symbol perspective, the performance data for the eight paper symbols show that the rules were learned and applied to all the symbols except the NDB. Classifying the NDB symbol as fly-by/fly-over or compulsory/on-request was difficult because the distinguishing feature, i.e., a circle surrounding the symbol or the fill of the symbol, may not have been salient. The circle distinguishing fly-over from fly-by may have been difficult to see given its proximity to the base symbol and because the global shape of the symbol did not change. Similarly, the fill of the symbol distinguishing it as compulsory or on-request may have been difficult to see because the area filled in the center of the symbol was small. Additionally, the resolution with which the NDB symbol was drawn may have been a factor; the NDB symbol used in the study is typically a paper-based symbol, and the resolution with which the symbol was drawn may not have been sufficient for the distinguishing features to be salient.

Classification of the ICAO recommended electronic symbols as being fly-by/fly-over was also problematic. Since a circle surrounds the ICAO electronic symbols, pilots tended to classify them as being fly-over rather than fly-by. The data speaks to the strength of the circle rule distinguishing fly-over waypoints from fly-by waypoints but also points to the potential for confusion if circles are used in the construction of symbols and are not intended to distinguish between fly-by and fly-over waypoints.

Finally, the data shows that the rules were easier to apply when more information was provided, i.e., when pilots were given a detailed legend that explicitly described the rule and provided many illustrative examples. Participants shown the detailed legend rather than the minimal legend were (1) slightly more accurate and significantly more confident in classifying symbols as fly-by or fly-over and (2) significantly more accurate, more confident, and faster in classifying symbols as ground-based or RNAV. There was no significant benefit to the presentation of a detailed legend for distinguishing compulsory vs. on-request symbols, suggesting that this symbol-feature rule was particularly easy to grasp with minimal information. While there are usually benefits for having more detailed information, it is important to note, however, that participants also stated that they generally do not look at legends before using charts. Thus, the results imply that during training, it may be more effective to show more illustrative examples of the rules and call pilots' attention to the legend where the rules are explicitly stated.

4 CONSIDERATIONS AND STEPS FOR EVALUATING SYMBOLS

The considerations in designing the experiments described in this report and the methods used may be useful to manufacturers and researchers interested in addressing in-depth issues in symbol design or in evaluating new or existing symbology. Symbols should be designed so that they are easy to find and identify and convey the information encoded without interfering with the interpretation of other symbols. The design and selection of symbols should consider the range of functions for which the symbols will be used. Existing symbols may need to be re-evaluated when new technology or displays are introduced or when it is determined that the current symbology is not sufficient (e.g., when information is being transferred from one medium to another such as transitioning from paper charts to electronic displays).

The two sections below summarize considerations and steps for designing and evaluating symbos. Section 4.1 summarizes considerations for symbol design. Section 4.2 summarizes the steps for designing an experiment; these steps may be useful for FAA evaluators and manufacturers interested in conducting their own evaluations. Additional guidance for testing symbols can be found in SAE ARP 4155, *Human Interface Design Methodology for Integrated Display Symbology* [7].

4.1 Symbol Considerations

1. *Ensure that the symbol is legible.* A symbol is legible if its shape is clear and perceptible. Legibility is dependent upon two factors: (a) the qualities of the display and (b) the symbol size.

    a) How a symbol looks on a display will vary depending on the physical qualities of the display. Display contrast, resolution, and size will influence a symbol's appearance. Legibility of the symbols should be examined under a variety of viewing angles, distances, and lighting conditions.

    b) The minimum size at which a symbol is presented must preserve the key features that define it. Not surprisingly, large symbols are easier to detect and discriminate than smaller symbols, but when display space is limited, a symbol should not be too large. When designing and evaluating symbols, it will be important to consider whether a minimum size should be specified, and how this minimum size may be influenced by characteristics of the display technology.

2. *Ensure that the symbol is distinctive.* A symbol is distinctive if it is easy to discriminate from other symbols, even if it differs from other symbols by only one feature. Symbols should be designed to have a basic shape or characteristic that can be recognized and easily identified by users. For example, a unique global shape will make identification easier and reduce the potential for confusion. In some cases, manufacturers may add details to the design of the base symbol so that they can create their own look and feel. If enhancements are added, it will be important to determine if those enhancements add confusion or adversely impact the recognition time or rate of the symbol. Additionally, it may be valuable to determine if these enhancements in the details of a symbol's design become critical to the recognizability of the symbol.

    The distinctiveness of a symbol should be evaluated not only within a symbol set but also across symbol sets to ensure that the symbols are consistent within and across the flight deck, including with other displays and charts (paper or electronic) and to ensure that the symbol is not confusable with other symbols or coding conventions used in the flight deck. The issue of consistency in symbol presentation and compatibility across electronic systems will be more important as electronic charts replace paper charts in the future.

3. *Ensure that the symbol is salient.* The saliency of a symbol refers to prominent it is or how much it stands out in a given context, e.g., within a cluttered background. Search for a symbol is heavily influenced by the total number of items shown on the display (i.e., global density) as well as the number of items proximate to the target symbol (i.e., the local density). As more information is depicted on the display (i.e., the global density) or as the number of items in close proximity to the

target symbol (i.e., the local density) increases, the target symbol becomes more difficult to find. Since local and global density vary widely from one chart to another, a symbol that may be easy to find in one context may not be in another.

Search for a symbol may be facilitated if a symbol can be designed to be more salient than other items. One way to increase the saliency of a symbol within a display is to manipulate features of that symbol, e.g., to make the symbol bold, to use color, or to use intensity differences. It is important that the use of color is considered carefully as excessive use can actually increase response time. Additionally, any use of color should be consistent with the flight deck color philosophy and should not interfere with other color-coding conventions in the flight deck.

4. *Ensure that the symbol is interpretable.* The meaning of a symbol may be easily identifiable within the chart because context clues (e.g., the location of the symbol on a chart or frequency information that accompanies the symbol) can be used to determine what the symbol represents, even if the meaning is not obvious. The meaning of a symbol should be measured without context clues to discern the meaning conveyed by the symbol itself. It may also be of interest to evaluate the symbol's meaning when some context is added, e.g., frequency information that can indicate to the pilot the symbol type. A final step would be to evaluate the meaning of the symbol in the full context of the chart.

4.2 Steps for Evaluating Symbols

1. *Collect symbols that are recommended and/or currently in use.* Use of standard symbology simplifies training and can reduce the chance that a symbol is misinterpreted. As many symbols as possible should be collected. Symbol sets can be found in ICAO *Annex 4* [2], the NACO *Aeronautical Chart User's Guide, 6th Edition* [4], and SAE ARP 5289 [6].

2. *Develop new symbols (if needed).* If current symbols are not sufficient, then new symbols that are distinctive and recognizable may be designed. This may be the case if new functions are introduced that require new symbols. In developing these new symbols, it will be important to consider pilot familiarity with charts (e.g., conventions for paper charts) and the potential training impact of introducing new symbols. Additionally, new symbols should have a unique global shape that distinguishes it from existing symbols; the design of new symbols that differ only slightly from current symbols should be avoided. For example, a new symbol that is a rotated version of an existing symbol may not be sufficiently distinctive. Similarity of symbols increases the potential for confusion and the chance for misinterpretation, particularly as workload increases.

   It is important to note that an excessive number of symbols should not be created. As the number of symbols in a symbol set increases, the time required to identify each symbol may also increase. Additionally, when new symbols are created, each symbol in a symbol set should be re-evaluated to ensure that each one can be recognized independently and within the full set.

3. *Recruit participants for the evaluation.* Ideally, participants evaluating symbols should be representative of the target end users, i.e., a range of the pilots representing the experience and ratings that are expected to use the equipment. Non-pilots may provide valuable initial input for some types of evaluations, e.g., if the purpose of the evaluation is to consider whether a symbol shape is distinctive from other shapes.

   Note that participants' background and training may influence how new symbols are interpreted and which symbols are easiest to recognize and use. Background information such as the type of charts pilots use, the type of aircraft flown, the type of route flown (e.g., domestic or international), and familiarity with FMS, moving map, or EFB technologies may be collected in order to determine if these factors influence the results. It is important to test pilots with a range of backgrounds in order to identify potential issues with they symbols early in the design process. See Appendix B for a sample background questionnaire.

4. *Evaluate symbols*. There are many tasks that could be used in evaluating symbols. For example, one could be asked to find symbols in the context of a chart in order to evaluate how easy a symbol is to find or one could be shown a symbol and asked to identify it. The task selected will determine the dependent variable to be measured. Dependent variables may be the time to find a symbol or the accuracy in identifying a symbol. In setting up an evaluation of the new symbols, it may be useful to create *foils*, i.e., shapes that are not currently in use, may be useful for inclusion in any study. In the experiments reported here, the foils provided verification that (1) pilots indeed had a stereotype for different symbol types and did not simply consider all shapes to be representative and that (2) pilots could generalize and apply symbol-feature rules used to code symbols.

5. *Determine symbol acceptability*. Results from the evaluation should be tabulated to measure the acceptability of the test symbols. A symbol is acceptable if the rate of recognition meets a criterion level of acceptability. Ideally, the criterion level of acceptability is defined through formal data analyses but it may also be defined by the experimenter, e.g., by achieving a mean accuracy recognition rate or response time set at a pre-defined level.

5 SUMMARY AND CONCLUSION

This report describes two experiments conducted to better understand the limits in variation for the design of navigation symbols. Experiment 1 addressed the issue of symbol stereotypes and whether there are key features of symbols necessary for recognition. Experiment 2 evaluated symbol-feature rules for coding symbols to determine if pilots could learn and apply them. The rules illustrate a logical, consistent way to design symbols to distinguish between a fly-over or fly-by waypoint, a compulsory or on-request reporting point, or a ground-based or RNAV fix.

Pilots' ratings in Experiment 1 identified representative shapes for seven of the eight navigation symbols, despite variations in the size, color, and fill of the test symbol shapes. The representative shapes were presented in Table 20 (see page 16). No representative shape was identified for a stand-alone DME in the results; pilots' comments indicated that they are more familiar with a DME presented in conjunction with another symbol.

The results of Experiment 2 indicated that pilots were generally able to learn and apply the symbol-feature rules, although the rules were easier to apply when more information was provided, e.g., in a detailed legend that explicitly described the rule and depicted many examples. Pilot comments indicated, however, that they typically do not read the legend before looking at their charts so there may be training implications in the adoption of these rules.

The results also identified considerations for symbol design in implementing the symbol-feature rules. In particular, the legibility of the NDB symbol when shown electronically affected the application of the compulsory/on-request rule and fly-by/fly-over rule. Because the NDB symbol is drawn with a high level of detail, the fill and circle features may not have been salient. Additionally, the application of the fly-over/fly-by rule to the ICAO electronic symbols was not intuitive. Many pilots classified these symbols to be fly-over rather than fly-by because the recommended ICAO electronic symbols are surrounded by a compass rose, i.e., a circle. This finding highlights the dominance of the circle rule because pilots categorized any symbol that was surrounded by a circular shape as fly-over. Thus, symbols that looked like a circle, but were not meant to represent fly-over waypoints, may be misinterpreted. The results show that key features used to convey information about the symbol (e.g., fill and presence/absence of a circle) must be salient and unambiguous when applied to all symbols and must be distinct from global features used to distinguish symbol classes.

These findings contribute towards the development of recommendations to FAA, industry, and ICAO regarding electronic symbology for navigation information. The results of the studies are planned to be used in industry efforts to develop recommended best practices for electronic symbols (e.g., in updating the recommended symbol set in ARP 5289 [6]). The studies described here provide input regarding what symbol shapes are easily recognizable and how those symbols may be modified. The next step will be to validate the symbol set proposed for an update to SAE ARP 5289 to ensure that the final recommended symbol set will be usable and recognizable. While the scope of this work addresses navigation symbology, the techniques used here are applicable for addressing other types of symbology as well.

6 REFERENCES

1. ICAO Aeronautical Information and Charts Study Group Meeting. Working paper. Concept for RNAV/Ground-Based Charting Symbol Consistency and Hierarchy. 3-5 June, 2004. Montreal, Canada.

2. International Civil Aviation Organization (ICAO) Annex 4. *Aeronautical Charts*, Annex 4 to the Convention on International Civil Aviation, 10th edition, July 2001.

3. International Organization for Standardization (ISO). (2001). *Graphical Symbols – Test Method for Judged Comprehensibility and for Comprehension, 2nd Edition*. ISO 9186. ISO: Geneva, Switzerland.

4. National Aeronautical Charting Office (NACO). (2004). *Aeronautical Chart User's Guide, 6th Edition*. Federal Aviation Administration: Washington, DC.

5. RTCA. (2003). *Minimum Operational Performance Standards for the Depiction of Navigational Information on Electronic Maps*. DO-257A. RTCA: Washington, D.C.

6. Society of Automotive Engineers (SAE). (1997). *Electronic Aeronautical Symbols*, ARP 5289. Society of Automotive Engineers: Warrendale, PA.

7. Society of Automotive Engineers (SAE). (1997). *Human Interface Design Methodology for Integrated Display Symbology*, ARP 4155. Society of Automotive Engineers: Warrendale, PA.

8. Yeh, M. and Chandra, D.C. (2003). An approach for designing flight symbology. Poster presented at the *HCI-Aero Conference*, 29 September – 1 October, 2003, Toulouse, France.

9. Yeh, M. and Chandra, D.C. (2004). Issues in Symbol Design for Electronic Displays of Navigation Information. *Proceedings of the 23rd Digital Avionics Systems Conference*. 24-28 October 2004, Salt Lake City, Utah.

APPENDIX A: BACKGROUND QUESTIONNAIRE

Age _____

Gender   Male   Female

Flight Hours   Total _____   Average (per month) _____
                                              Last month _____

Instrument Time Total _____   Average (per month) _____
                                              Last month _____

Which manufacturer provides the charts that you use most? How long?
_____

Do you use charts from other manufacturers regularly? Which? How long?
_____
_____
_____

Ratings and Certificates: Please check the ratings and certificates that you have.

Airline Transport _____   **Instructor**
Commercial _____   Certified Flight Instructor _____
Glider _____   Certified Instrument Instructor _____
Private _____   Multi Engine Instructor _____
Recreational _____   Ground Instructor _____
Rotorcraft _____
Student _____

Instrument _____
Single Engine _____
Multi Engine _____

Please list other ratings that you hold: **List aircraft you have flown in the past.** Please indicate which aircraft you fly most frequently.

Type Rating

_____   _____
_____   _____
_____   _____
_____   _____
_____   _____
_____   _____

-- OVER --

Flight Experience: Please check the type(s) of flying that you do:

Instruction   IFR ____   VFR ____

Recreational   IFR ____   VFR ____

Business ____

Corporate ____

Air transport ____

Military ____

Do you have experience with the following:

- **Glass cockpit?**   Yes   No
- **FMS?**   Yes   No
- **moving map displays?**   Yes   No
- **electronic flight bags?**   Yes   No

DRAFT

APPENDIX B: SYMBOL STEREOTYPES QUESTIONNAIRE

INSTRUCTIONS

The purpose of this study is to understand what key features are necessary for symbols to be recognized. This experiment addresses eight symbols: DME, fix, NDB, TACAN, VOR, VOR/DME, VORTAC, and waypoint. For each symbol type, you will be shown a set of symbol shapes and asked to cross out those shapes that are NOT representative of the symbol type. You will then be asked to explain how you classified the symbols by drawing a shape(s) that you consider to be representative of the symbol type and providing a text description of the rule you used in crossing out the shapes.

A practice set is shown below.

PRACTICE: AIRPORT

Based on your knowledge of charts and navigation displays, decide whether the symbol would represent an AIRPORT or not if you saw it on a chart or navigation display.

Cross out the symbol(s) that you would NOT consider to be an AIRPORT.

Draw the shape(s) that you feel is most representative of an AIRPORT symbol. Write the rule you used to decide if the symbols above were AIRPORT symbols or not.

Symbol Shape	Rule

Jeppesen symbols have been reproduced with permission and are copyrighted by Jeppesen Sanderson, Inc.
Some symbols are reduced or use different colors for illustrative purposes.

DME

Based on your knowledge of charts and navigation displays, decide whether the symbol would represent a DME or not if you saw it on a chart or navigation display.

Cross out the symbol(s) that you would NOT consider to be a DME.

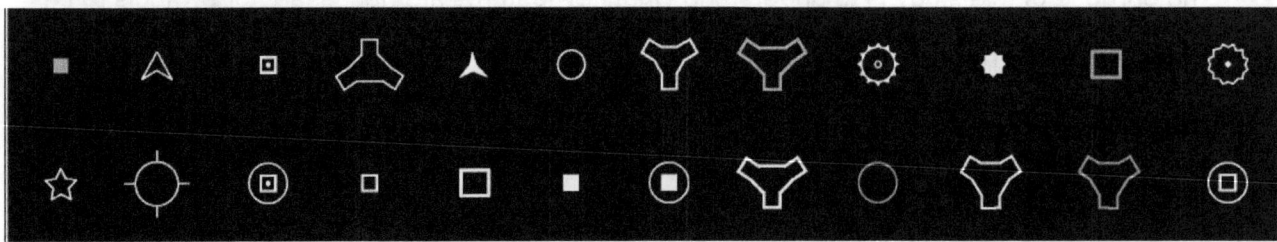

Draw the shape(s) that you feel is most representative of a DME symbol. Write the rule you used to decide if the symbols above were DME symbols or not.

Symbol Shape	Rule

TACAN

Based on your knowledge of charts and navigation displays, decide whether the symbol would represent a TACAN or not if you saw it on a chart or navigation display.

Cross out the symbol(s) that you would NOT consider to be a TACAN.

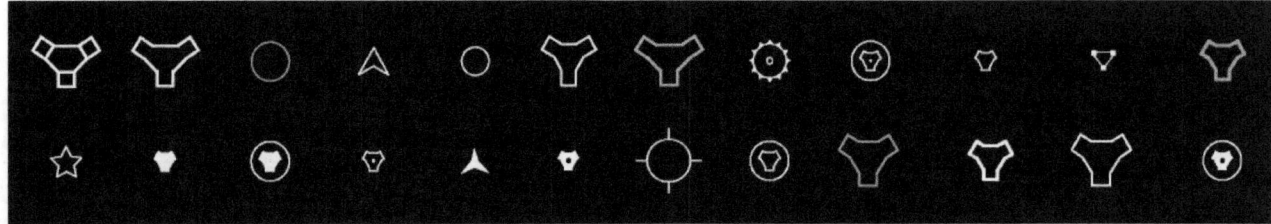

Draw the shape(s) that you feel is most representative of a TACAN symbol. Write the rule you used to decide if the symbols above were TACAN symbols or not.

Symbol Shape	Rule

VOR

Based on your knowledge of charts and navigation displays, decide whether the symbol would represent a VOR or not if you saw it on a chart or navigation display.

Cross out the symbol(s) that you would NOT consider to be a VOR.

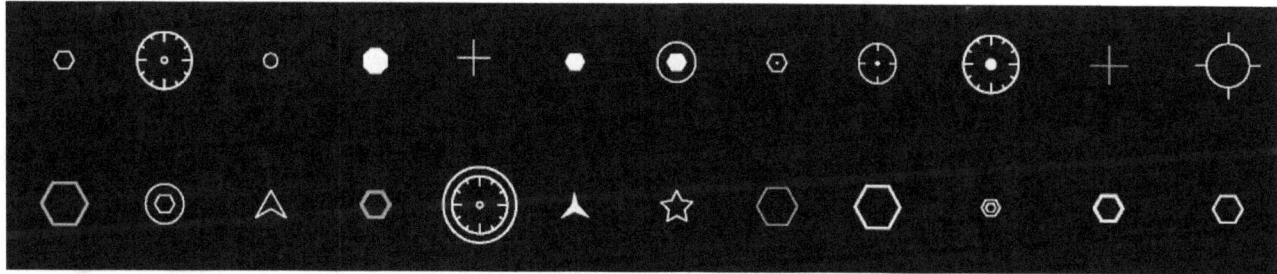

Draw the shape(s) that you feel is most representative of a VOR symbol. Write the rule you used to decide if the symbols above were VOR symbols or not.

Symbol Shape	Rule

VORTAC

Based on your knowledge of charts and navigation displays, decide whether the symbol would represent a VORTAC or not if you saw it on a chart or navigation display.

Cross out the symbol(s) that you would NOT consider to be a VORTAC.

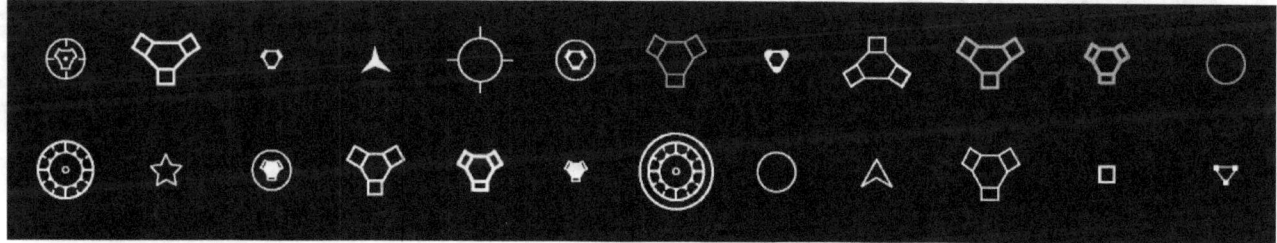

Draw the shape(s) that you feel is most representative of a VORTAC symbol. Write the rule you used to decide if the symbols above were VORTAC symbols or not.

Symbol Shape	Rule

Jeppesen symbols have been reproduced with permission and are copyrighted by Jeppesen Sanderson, Inc.
Some symbols are reduced or use different colors for illustrative purposes.

VOR/DME

Based on your knowledge of charts and navigation displays, decide whether the symbol would represent a VOR/DME or not if you saw it on a chart or navigation display.

Cross out the symbol(s) that you would NOT consider to be a VOR/DME.

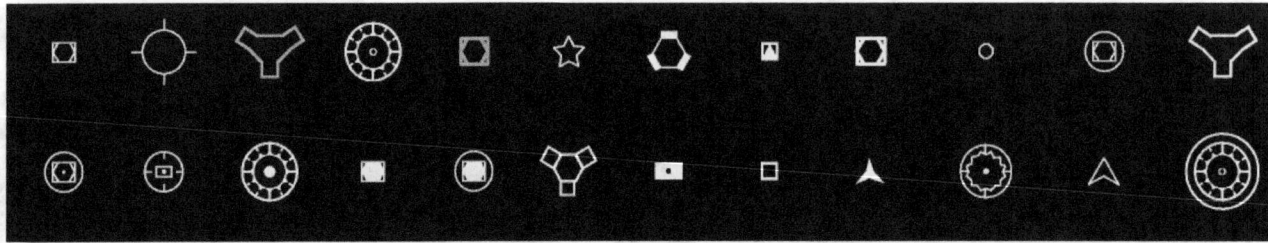

Draw the shape(s) that you feel is most representative of a VOR/DME symbol. Write the rule you used to decide if the symbols above were VOR/DME symbols or not.

Symbol Shape	Rule

Waypoint

Based on your knowledge of charts and navigation displays, decide whether the symbol would represent a WAYPOINT or not if you saw it on a chart or navigation display.

Cross out the symbol(s) that you would NOT consider to be a WAYPOINT.

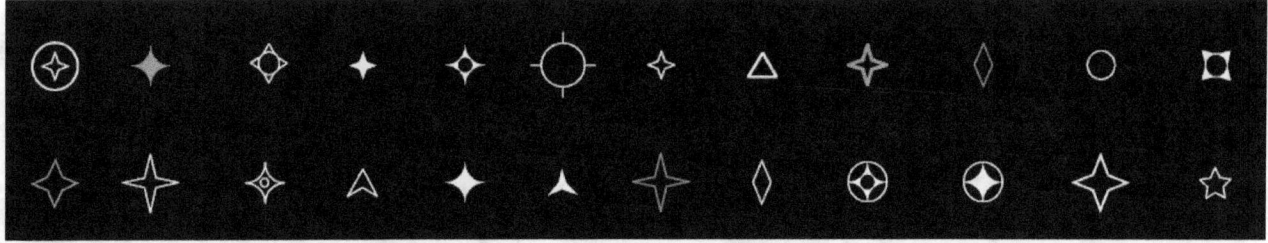

Draw the shape(s) that you feel is most representative of a WAYPOINT symbol. Write the rule you used to decide if the symbols above were WAYPOINT symbols or not.

Symbol Shape	Rule

Jeppesen symbols have been reproduced with permission and are copyrighted by Jeppesen Sanderson, Inc.
Some symbols are reduced or use different colors for illustrative purposes.

Fix

Based on your knowledge of charts and navigation displays, decide whether the symbol would represent a FIX or not if you saw it on a chart or navigation display.

Cross out the symbol(s) that you would NOT consider to be a FIX.

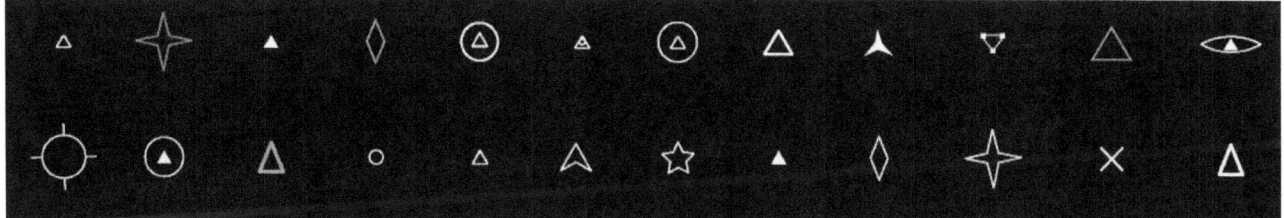

Draw the shape(s) that you feel is most representative of a FIX symbol. Write the rule you used to decide if the symbols above were FIX symbols or not.

Symbol Shape	Rule

NDB

Based on your knowledge of charts and navigation displays, decide whether the symbol would represent an NDB or not if you saw it on a chart or navigation display.

Cross out the symbol(s) that you would NOT consider to be an NDB.

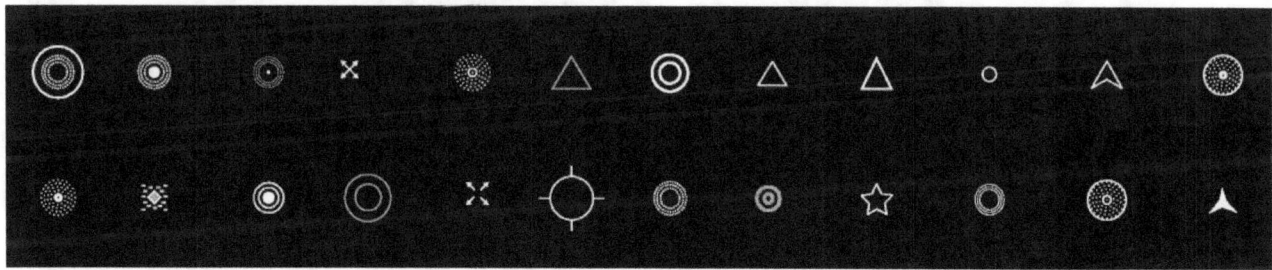

Draw the shape(s) that you feel is most representative of an NDB symbol. Write the rule you used to decide if the symbols above were NDB symbols or not.

Symbol Shape	Rule